ICE AGE EXTINCTIONS, A NEW THEORY

ICE AGE EXTINCTIONS, A NEW THEORY

Explains Megafaunal, Neanderthal, Hobbit extinctions and Geomagnetic Reversals

John Stojanowski

Pangea Publications, LLC
Staten Island, New York

Cover image of mastodon from exhibit at the Staten Island Museum at Snug Harbor, Building A, 1000 Richmond Terrace, Staten Island, N.Y.

ISBN: 978-0-9819221-5-7

Graphics created or enhanced with CoPlot

10 9 8 7 6 5 4 3 2 1

CONTENTS

INTRODUCTION

The controversy regarding the cause, or causes, of megafaunal extinctions that occurred during the Pleistocene Epoch (2.6 million yearBP to 10,000 yearBP) and more specifically over the last 100,000 years continues to this day. The three major contenders for the causation of the extinctions are:

1) climate change, the "Overchill hypothesis";
2) man killed them in a great slaughter, the "Overkill hypothesis"; or
3) man's diseases killed them off, the "Overill hypothesis"

Based on recent research (published in 2015) the ascendant hypothesis is that the abrupt warming periods known as "interstadials" that alternated with freezing glacial periods ("stadials") were responsible for the extinctions. The exact mechanism(s) causing the extinctions during the interstadials has not been proffered. Leading paleogeneticist Alan Cooper concluded, based on this recent research:

"We can see the relationship between the warming periods and extinctions but can't tell whether it's the warming or the pace of change. It's one of the two."

At the end of the most recent ice age, the time of terminal-Pleistocene megafauna extinctions, woolly mammoths lived alongside woolly rhinoceroses, cave lions, cave bears, saber-tooth tigers, dire wolves, ground sloths, the "Irish elk," various types of horses, several types of bison and many other oversized fauna. What is significant is that megafauna extinctions didn't occur simultaneously around the globe during the last 50,000 years of the Pleistocene.

This book introduces a new theory to the mix. I am the author of the theory and I use the word "theory" in place of the word "hypothesis" because I believe my theory is the only one that can accurately account for the extinctions of the megafauna, Neanderthals and Hobbit (i.e., Homo floresiensis.) It is supported strongly, but indirectly, by the recent research verifying the interstadial/extinction linkage as well as many other factors that will be described henceforth.

My theory, **The Gravity Theory of Mass Extinction**[19] (**GTME**), is a little over eleven years old. It was written to initially explain the emergence of the megafauna, e.g., dinosaurs, that existed over sixty five million years ago as well as the reason for their extinction. The theory was then expanded to explain the primary cause of the Big Five Mass Extinctions as well as others. After studying the Quaternary megafaunal extinctions I realized that the same mechanism for extinction, namely changes in surface gravity on the Earth, was the reason the ice age megafauna attained their extreme, above normal size as well as the reason why they became extinct. Although the mechanism causing the surface gravity change is somewhat different between the two periods, the effect on life forms was similar. Also, the case will be made that the Neanderthals and Hobbit became extinct primarily from the same effect that extirpated the megafauna, namely changes in surface gravity. Note that said change in surface gravity occurred on parts of the Earth's surface at different times, not the entire surface at the same time, i.e., there was a gravitational gradient around the globe and the regions with lower than current surface gravity levels changed over time as did the regions with higher than current surface gravity. Also, the regions with the lowest surface gravity were antipodal to the regions with the highest surface gravity.

As improbable as it may seem the GTME can also explain why the magnetic field of the Earth periodically reverses; the northern magnetic pole switches position with the southern magnetic pole. The GTME can also explain why flood basalt volcanism occurs on the Earth's surface.

This book will explain the Quaternary megafaunal extinctions as well as the above mentioned phenomena based on **The Gravity Theory of Mass Extinction (GTME).**

CHAPTER 1

QUATERNARY EXTINCTIONS-THE HYPOTHESES

1.1 HUMAN HUNTING

The theory that human hunting was the primary cause of the megafaunal extinctions (**the Overkill hypothesis**), until recently, had a lot of momentum. It was believed that in every one of the megafaunal extinction periods humans had arrived immediately prior to or at the same time as the extinctions and the extinctions were caused by human activity, especially hunting. Although this would appear to be true for the later Quaternary extinctions, it is not true of the earlier megafaunal extinctions, which will be explained in a following section on climate change.

The leading advocate for the human hunting hypothesis was Paul S. Martin, a geoscientist who has extensively researched and written about the ice age megafaunal extinctions. His "Blitzkrieg" hypothesis regarding human hunting is well known. In the book *Quaternary Extinctions*[4] he lists 8 attributes of the late Pleistocene extinctions:

1. Large mammals were decimated.
2. Continental rats survived; island rats did not.
3. Large mammals survived best in Africa.
4. Extinctions could be sudden.
5. Regional extinctions were diachronous (i.e., not happening at the same time).
6. Extinctions occurred without replacement.
7. Extinctions followed man's footsteps.
8. The archeology of extinction is obscure.

In the past, human hunting was considered the most likely cause of the late Pleistocene extinctions because humans appeared in the regions of extinction near the time of the extinctions.

However, there are issues with this hypothesis. In North America at least 3 dozen genera became extinct in the late Pleistocene but only 3 have been found to be associated with early human remains, namely camels, horses and mammoths. The assertion that "extinctions followed man's footsteps" has been challenged in references to places such as Australia.

1.2 CLIMATE CHANGE

Extinctions due to climate change (the Overchill hypothesis), specifically the cooling of the Earth's temperature, is one of the primary hypotheses seeking to explain the periodic extinction of megafauna during the last 60,000 years and much earlier. During this period there have been cycles of warming, known as "interstadials", alternating with cooling periods ("stadials"). It was believed that the cooling periods had a negative effect on plant life causing the extinction of megafauna that couldn't adapt to the changes in food sources. And, it was thought that the megafauna were not able to migrate to regions not severely affected by the cooling. Gestation time, diet, temperature tolerance and other ecological factors were believed to have played a part in the extinctions during these periods.

In addition, other scientists believe that when the change to warmer periods commenced, deep changes in seasonality would affect the ability of the megafauna to successfully reproduce. Their offspring would have to be born at a certain time of year to be ready to face the harsh winters. It was believed that the megafauna couldn't evolve fast enough to alter their reproductive behavior. The scientists who believe in the stadial climate change causation of extinction have concluded that the periods of severe cold weather were the primary cause of the extinctions. Without being able to specifically provide a consistent climate change explanation for megafaunal extinctions at different times and on different continents the supporters of the climate change hypothesis for the megafaunal extinctions have had a difficult time convincing the scientific community that they are correct. The fact that previous cold periods of climate change didn't seem to account for the extinctions weakened their support for climate induced extinction.

Recently, with the help of DNA and radiocarbon data, scientists have been able to accurately correlate the warming periods of the cold/warm cycles with the extinction of megafauna during the past 60,000 years. The data contradicts the belief that the colder periods had the greater impact on megafauna and were responsible for the extinctions. Surprisingly, it was within the warm periods, the "interstadials", that the extinctions occurred. The duration of the warming periods were much briefer than the colder periods. The research[9] was undertaken by a group from the University of Adelaide in Australia led by Alan Cooper, an ancient DNA paleogeneticist researcher.

The DNA evidence unveiled a surprise. During several of the extinction periods, certain megafauna that went extinct were eventually replaced by similar megafauna making it appear that there were no extinctions. This revelation has weakened the human hunting hypothesis which was previously supported by the premise that there were few extinctions during prior cycles of warm/cold intervals and the extinctions always followed the migration of humans.

The researchers emphasized the point that the extinctions occurred during the 60,000 year period at times **regardless of whether humans were present**, casting convincing doubt on the "overkill" hypothesis. As an example, they point to the Short-faced bear of North America that went extinct before humans appeared. The researchers also stated:
"Several taxa (e.g., mammoth) go extinct on the mainland of Eurasia considerably later than that of the New World, despite a much longer exposure to human hunting."

They concluded that the rapid rise in temperature had a much more negative effect than the adjacent longer duration cooling periods as well as the presence of humans. Chris Turney of the

University of New South Wales in Sydney, Australia stated:
"What we found, which we were staggered by: no matter how we analyzed the data, abrupt warming drove the extinctions and the replacements."

The researchers concede that humans may have had a less important role in the extinctions as the weakened megafauna were forced into smaller habitable areas. Also, they suggest that farming by humans 12,000 years ago may have limited the land area accessible to the megafauna, hastening their demise.

Cooper stated:
"We can see the relationship between the warming periods and the extinctions but we can't tell whether it's the warming or the pace of the change."

The theory written by the author of this book will explain why the extinctions occurred during the warming periods.

1.3 DISEASE

The Hyperdisease hypothesis (**the Overill hypothesis**) is based on the premise that the early human arrival in North and South America unleashed the spread of disease to the largest mammals living there. Experts have studied this possibility and have rejected this hypothesis.

The Hyperdisease hypothesis asserts that smaller mammals were not affected because they had a short gestation period and a larger population size. Scientists state that there are at least four requirements for disease to be responsible for extinction:

1. The disease must be able to sustain itself when no susceptible mammals are available to infect.

2. The disease must have a high infection rate, able to infect all mammals regardless of age or sex.

3. The disease must be extremely lethal with a mortality rate on the order of 50-75%.

4. Most important, the disease must be able to infect multiple host species without affecting humans to any significant degree, i.e., not able to create an epidemic.

Domesticated dogs (and chickens) that accompanied the early humans have been discarded as the disease source because the megafauna extinctions in Australia occurred long before these animals were introduced there. Dogs didn't arrive in Australia until about 35,000 years after humans had reached that continent which is 30,000 years after the megafaunal extinctions had ended.

Ross MacPhee of the American Museum of Natural History along with Preston Marx of the Tulane University of Louisiana are the chief proponents of the Hyperdisease hypothesis. They assert that newly emerged diseases, as opposed to long established ones, can have lethal consequences when they jump the species barrier. Critics of this hypothesis point out that not only mammals were wiped out but also birds and reptiles, such as the extinct giant monitor lizard Varanus priscus from Australia. They feel that it is unlikely that a single infectious disease could eliminate such a wide diversity of fauna.

FIG. 2-1 Cave Bear, American Museum of Natural History

CHAPTER 2
THE NEW THEORY

The **Gravity Theory of Mass Extinction** (**GTME**) was written by the author of this book to explain the emergence of the megafauna of the Paleozoic and Mesozoic Eras, especially some dinosaurs, as well as the reason for their extinction. In the process of developing the theory it became apparent that the primary cause of all major mass extinctions were linked to the same phenomena that allowed some ancient megafauna to attain their unnaturally large size. That phenomena was the periodic change of the Earth's surface gravity. As of the publication of this book, the theory has not been officially recognized as the cause of the abnormal size of ancient megafauna nor the reason for their mass extinction although it has received little attention from the scientific community. The author of this book believes that changes in surface gravity during the Quaternary Epoch were responsible for the Ice Age megafauna as well as their periodic extinction and that the publication of this book will, hopefully, confirm the validity of the GTME.

2.1 THE GRAVITY THEORY OF MASS EXTINCTION- AN OVERVIEW

This theory is over eleven years old and is described more fully in a book[19] with the same name and on various websites. The basis of this theory is that, under certain conditions, one or more of the Earth's core elements, including the inner core, outer core and the densest part of the lower mantle surrounding the cores, can and have moved away from their current central location within the Earth's equatorial plane. While it may sound counterintuitive at first, there is ample evidence that this has happened. Once any one of the three of the core elements moves away from Earth-centricity, by definition, surface gravity will be different at various points on the globe because the distance between the altered center of mass of the Earth and different points on the Earth's surface will also be different. This gravitational dependence on distance is explained by Isaac Newton's Universal Law of Gravitation.

The scientific explanation for the offsetting of one or more of the core elements is based upon a scientific principle known as the Law of Conservation of Angular Momentum. One example frequently used to explain this concept is the ice skater that goes into a spin and moves its arms away from its body and then moves them tightly to the side. The result is a slowing of angular (i.e., rotational) velocity followed by a rapid increase in angular velocity. What this demonstrates is that the distance of the skater's movable mass (i.e., arms) from the skater's axis of rotation, which is a vertical line passing through the center of the skater's body, determines the change in angular velocity of the skater after the skater initiates the spinning. No external forces are needed to cause the change in rotational velocity of the skater, thereby conserving angular momentum.

Let's substitute the spinning Earth for the ice skater. The Law of Conservation of Angular Momentum must still apply. Let's replace the skater's arms with a supercontinent that has its center of mass at a high southern latitude. We then observe the supercontinent's center of mass moving from a high southern latitude north toward the equator, due to plate tectonics. We know that this is something that actually happened approximately 300ma to 250ma with the supercontinent of Pangea and can be seen in a graph published in a research paper[1] entitled '*Plate tectonics may control geomagnetic reversal frequency*' by Petrelis, Besse and Valet in 2011 (see References). Based on this movement to a lower latitude, the supercontinent's center of mass moved further away from the Earth's axis of rotation just as the skater's center of mass moved away from its axis of rotation when its arms were extended outward. Therefore, we would expect the Earth's rotational velocity to decrease in a manner similar to that of the skater. Surprisingly, when Pangea's center of mass moved from a high southern latitude to a lower one the Earth's rotational velocity didn't change. Why not?

The Earth's rotational velocity didn't change because there was something else that compensated for the supercontinent's center of mass movement away from the Earth's axis of rotation as it moved to a lower southern latitude. That compensating factor was the movement of one or more of the Earth's core elements toward Earth-centricity and toward Pangea. Note that the Earth's core elements were already significantly displaced when Pangea's center of mass was at a high southern latitude. If one visualizes a skater with two arms of normal length in addition to two short massive sized arms the concept is easily understood. The skater starts spinning with the normal length arms close to the body and the shorter, heavier arms stretched out away from the body. The skater then slowly extends the normal arms away from the body as it pulls the short heavier arms toward its body. If synchronized properly, the skater's angular velocity will not change and neither will its angular momentum, which must be conserved. Therefore, the Earth's core elements acted in the same way the skater's short heavier arms did, i.e., moving closer to the axis of rotation in order to maintain the Earth's angular momentum as Pangea's center of mass moved to a lower southern latitude. No change in the Earth's rotational velocity occurred when Pangea's center of mass moved further away from the Earth's axis of rotation.

Note that the primary difference between the skater and the Earth in the above comparison is that the skater had a balanced symmetry with its two (or four) arms while the Earth, with the single supercontinent of Pangea did not. This would have caused a wobble of the Earth at certain times. In fact, scientists found that an unusual major wobble occurred during the late Cretaceous Period[12] after Pangea had broken apart and the continental remnants were moving primarily longitudinally toward their current location. The author of this book asserts that the bolide impactor that struck at Chicxulub, Mexico about 65ma was dislodged from its orbit by this wobble of the Earth. The major cause of the K-T extinctions was the increase in surface gravity, not the bolide impact nor the Deccan Traps volcanism although the latter may have magnified the extinction.

When Pangea moved to a higher northern latitude (starting about 252 to 250 Ma after its center of mass crossed the equator) surface gravity on Pangea started lowering as the core elements moved away from Earth-centricity and away from Pangea. The lowest surface gravity was in the equatorial region of Pangea because the distance between the shifted core elements and the surface of the Earth (on the half of the Earth that Pangea was on) was greatest in the equatorial region of Pangea. The lower surface gravity on Pangea would have caused the sea level near Pangea to be higher than the sea level antipodal to Pangea. This is illustrated in FIG. 2-2.

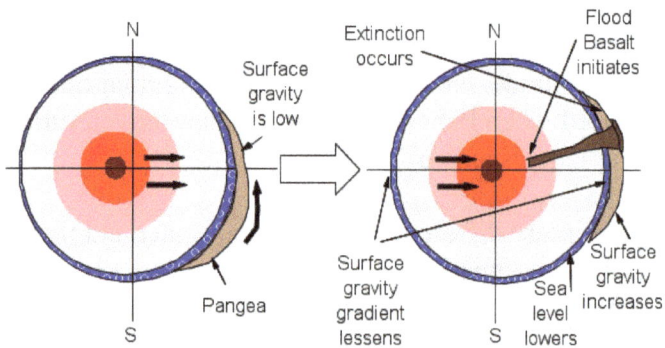

AS PANGEA BECOMES SYMMETRICAL TO EQUATOR, CORES MOVE, FLOOD BASALT ERUPTION INITIATES, SEA LEVEL LOWERS, SURFACE GRAVITY INCREASES ON PANGEA AND EXTINCTIONS OCCUR

FIG. 2-2 Effects of Earth's core movement toward Earth-centricity

The Gravity Theory Of Mass Extinction explains the emergence and massive size of the megafauna of the Phanerozoic and Mesozoic Eras as well as their extinctions as being due to an initial decrease in surface gravity followed by increase in surface gravity on Pangea, based on the above explanation. This cycle happened multiple times. The latitudinal movement of continental mass, e.g., the supercontinent of Pangea when it or its broken-apart remnants of the late Cretaceous moved to lower latitudes, was responsible for the shifted core elements moving back toward Earth-centricity. This movement increased surface gravity on Pangea to levels closer to, but never higher than, current surface gravity. But how does this relate to the Ice Age megafauna and their extinctions?

Describing The Gravity Theory Of Mass Extinction more generally, the basic concept is that when mass on the surface of the Earth moves so that there is a net latitudinal movement of mass to a higher latitude, one or more of the core elements will move radially away from Earth-centricity within the Earth's equatorial plane. This movement will change the Earth's surface gravity in a way that results in a gravitational gradient around the globe; lowest surface gravity at a point on the equator and highest surface gravity antipodal to that point.

During the Quaternary Epoch there was little latitudinal continent movement resulting from plate tectonic activity. Therefore, continental movement wasn't responsible for changes in the Earth's surface gravity as it was during the era of dinosaurs. However, we know there was massive movement of mass, in the form of water, to both polar regions where it accumulated in the form of ice during many periods within the Quaternary Epoch. Unlike the much earlier changes in surface gravity which were initiated by plate tectonics, i.e., due to continental mass moving latitudinally, the Quaternary megafauna attained their above normal size due to gravitational changes resulting from core element movement initiated by the transfer of ocean water to the polar regions. This is the concept of moving mass on the Earth closer to the axis of rotation described earlier with the example of the skater. The periodic melting of the polar ice during the interstadials transferred mass, in the form of water to lower latitudes, and therefore, away from the Earth's axis of rotation, resulting in core element movement back toward Earth-centricity increasing surface gravity in the regions that previously had lower surface gravity. This is why the research study mentioned in Chapter 1 found that the megafaunal extinctions occurred during the warming interstadial periods when polar ice was melting.

Therefore, the author of this book asserts that **the existence and periodic extinctions of the megafauna during the Quaternary Epoch were due primarily to surface gravity changes caused by the movement of mass, in the form of water, to and from the polar regions where it was transformed into ice.**

Because the Earth's warming periods, the interstadials, melted the ice delivering mass (in the form of water) to lower latitudes, it was responsible for the movement of core elements toward Earth-centricity causing an increase in surface gravity in areas that had lower surface gravity. Based upon the above description of the cause of surface gravitational changes during the Quaternary Epoch, it should be apparent that, at any instant in time, the longitudinal area experiencing the lowest surface gravity would be 180 degrees away from the area experiencing the highest surface gravity. In other words, there was a gravitational gradient, primarily longitudinal, around the globe. And, at any specific longitude there was also a smaller latitudinal gravitational gradient with the greatest change occurring in the equatorial region. This global gravitational gradient during the maximum extent of the last glacial maximum period (of about 26,000-22,000 yearBP) is illustrated in FIG. 2-3 (although the amount of polar ice is not shown; see FIG. 2-4 for a more accurate polar ice configuration).

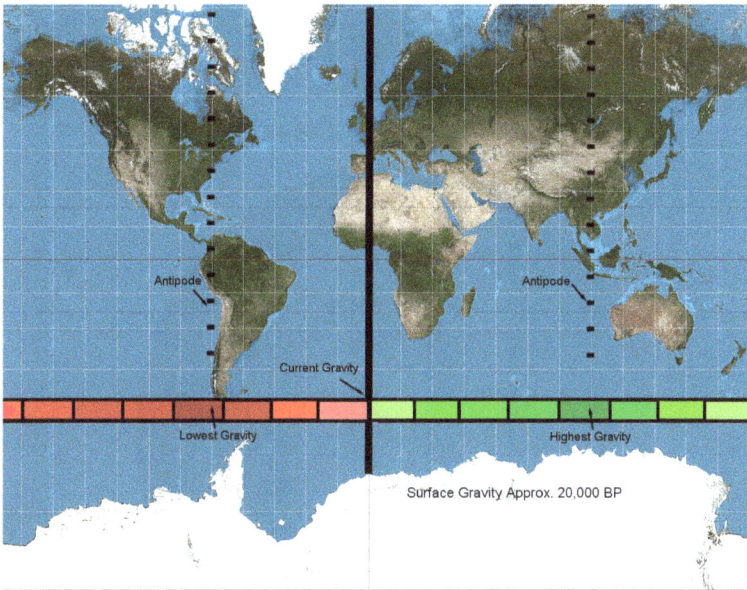

FIG. 2-3 Global gravitational gradient ~20,000 yearBP

From FIG. 2-3 **the longitudinal region that would be subject to lowest surface gravity is determined by the net asymmetrical distribution of polar ice in both polar regions combined, relative to the Earth's axis of rotation. The Earth's three core elements (from 1 to 2 to 3) will move radially in the Earth's equatorial plane in the opposite direction of the longitude containing the most ice mass.**

In other words, the land mass in the longitudinal region with the most polar ice will experience the lowest surface gravity because one or more of the core elements will move away radially from that longitude within the Earth's equatorial plane. And, the location of this region has changed throughout the entire Pleistocene Epoch.

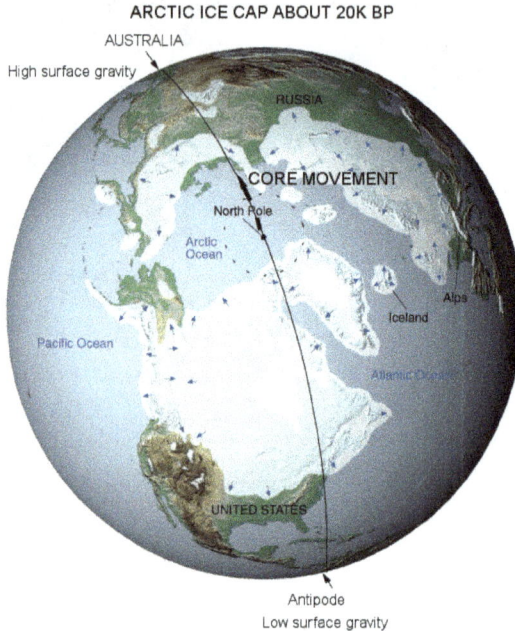

FIG. 2-4 Antipodal points with high and low surface gravity

Clearly, this means that the largest megafauna could not exist at antipodal positions on the Earth at the same time (see FIG. 2-4). And, due to the changing distribution of polar ice it can be deduced that megafaunal extinctions will have occurred at different times in different parts of the world. This assertion is supported by the chronology of late Quaternary extinctions (from Wikipedia):

1. **Australia- around 45,000 years ago**
2. **Tasmania- 41,000 years ago**
3. **Japan- 30,000 years ago**
4. **North America- 13,000 years ago**
5. **South America- 12,500 years ago**

6. Cyprus- 10,000 years ago
7. The Antilles- 6000 years ago
8. New Caledonia and nearby islands- 3000 years ago
9. Madagascar- 2000 years ago
10. New Zealand- 700 years ago
11. Mascarene- 400 years ago

Note that there is some disagreement about the timing of the above events.

Items 7-11 in the above list will not be addressed because these extinctions occurred after most of the polar ice cap had melted and are less likely to be the result of gravitational changes. Human hunting or climate change would most likely have been responsible.

The Climate Change hypothesis is on the right track when it associates the interstadial warming periods with the extinctions. However, it's not the local temperature change that's the culprit but the temperature change that causes the polar ice to melt. For example, Australia didn't experience a radical climate change approximately 45,000 yearBP when their megafauna became extinct. The "Overkill" human hunting advocates point this out when they try to discount local climate change as the cause of the Australian extinctions.

S. David Webb[13], a paleontologist, hypothesized that if the changes in climate resulting from glacial terminations causes land mammal extinctions then it would be expected that the extinctions would be diminished at lower latitudes. He cites extant tapirs, spectacled bears, giant tortoises, capybaras, peccaries and llamas to support his hypotheses. He stated: *"Nearly half of the groups that are counted as Rancholabrean (i.e., about 10,000 yearBP) extinctions at temperate latitudes have tropical survivors."*

 He attributes greater extremes in temperature and moisture availability along with plant and animal misadaptations during glacial termination periods for the megafauna extinctions.

The GTME would not attribute local climate changes as the primary factor in the extinctions. For example, when the glaciers melted surface gravity increased in a region that had lower surface gravity causing the megafauna in that region to deal with the increased energy requirements caused by the higher surface gravity. This energy requirement was much higher at higher latitudes due to the colder temperatures. Currently, the African elephant spends 80% of its time consuming food in order to survive. Even if an abundant quantity of food were available to it in northern latitudes it could not survive in those regions foraging for food because its energy requirement would be much higher than the available food supply could provide.

The GTME, if valid for the Pleistocene extinctions, requires that the western hemisphere containing North and South America to have had lower surface gravity immediately prior to the extinctions of roughly 13,000 to 11,000 yearBP. Likewise, the antipodal eastern hemisphere from approximately east India to western Australia would have had higher (than current) surface gravity at that time. Surface gravity strength in both hemispheres would be "gradientized" longitudinally as show in FIG. 2-3.

FIG. 2-5 displays the continental positions relative to the North Pole and the antipodal regions around the globe can be seen. As stated earlier, according to the GTME, the megafauna that went extinct could not exist at antipodal regions at the same time. For example, in Australia immediately prior to the period of megafaunal extinctions, of about 50,000 to 46,000 yearBP, there wouldn't have been megafauna in eastern South America at the same time.

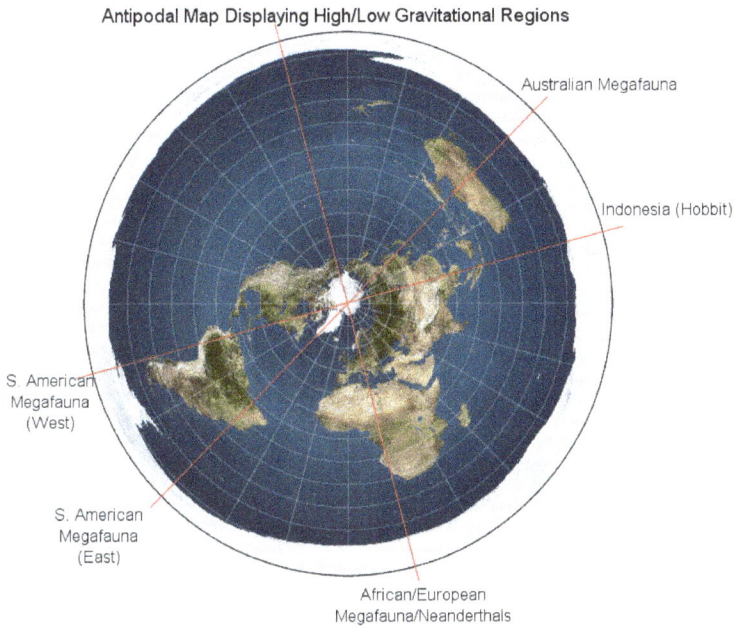

Antipodal Map Displaying High/Low Gravitational Regions

Australian Megafauna

Indonesia (Hobbit)

S. American Megafauna (West)

S. American Megafauna (East)

African/European Megafauna/Neanderthals

FIG. 2-5 Antipodal points around the globe

FIG. 2-6 Warren Mastodon 11,000 yearBP, American Museum of Natural History

CHAPTER 3
TIMING OF THE LATE QUATERNARY EXTINCTIONS

The approximate dates of the late Quaternary megafaunal extinctions were listed previously and are from Wikipedia. FIG. 3-1 illustrates the timing of these extinctions on a world map.

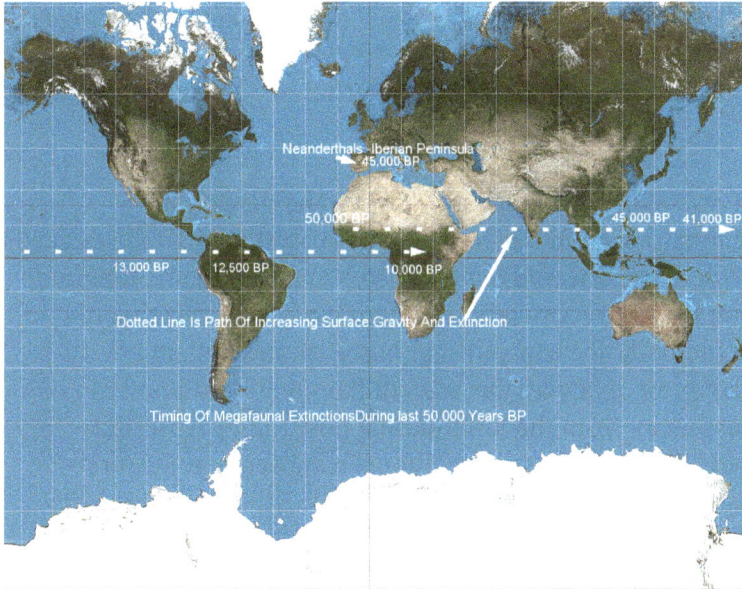

FIG. 3-1 Timing of Extinctions during last 50,000 yearBP

FIG. 3-1 illustrates an eastward progression of the extinctions from about 50,000 yearBP to about 10,000 yearBP. This pattern is consistent with a movement of increasing surface gravity that starts in Western Europe and moves gradually east over tens of thousands of years. But why would surface gravity gradually roll eastward in this manner?

As has been described above, according to the GTME, surface gravity is altered when a large amount of polar ice accumulates.

That accumulation must also be unsymmetrical relative to the Earth's axis of rotation. If it were possible to accumulate the polar ice symmetrically relative to the axis, it would not cause any of the Earth's core elements to move away from Earth-centricity and hence, no change in surface gravity would result. It would cause a minor increase in the Earth's rotational velocity. In fact, as this book was being written a scientific report confirming recent changes in rotational velocity caused by melting glaciers was published (*"Reconciling past changes in Earth's rotation with 20th century global sea-level rise: Resolving Munk's enigma."*) as reported in Science Advances 12/11/15.

Based on the above explanation, when the megafaunal extinctions started in Australia about 50,000 yearBP the largest mass of polar ice would have had to be near the same longitude as Australia. The eastward progression of extinctions would be due to the polar ice cap shifting, based on melting and redistribution, in a way that moved the greatest amount of polar ice mass away from the longitudinal position of Australia followed by the same process for each subsequent location where megafaunal extinctions took place. In other words, the polar ice mass in the polar longitudinal region in which Australia was located would be retreating, therefore the core elements started moving back towards Earth-centricity increasing surface gravity in that region. Note that the net change in distribution of polar ice involves the combined North and South polar ice caps, which complicates the matter. It appears that most of the polar ice redistribution occurred in the northern polar region.

A new study (in 2015) by archaeologist Todd Surovell from the University of Wyoming would appear to support Paul Martin's hypothesis that human hunting was the cause of the extinctions. Surovell's research[14], using radiocarbon dates from extinct mammals, found that decline of these megafauna correlated with the human migration across Alaska's Bering Strait through North America and into South America. He states that the corresponding decline in megafauna occurred 13,300-15,000 yearBP, 12,900-13,200 yearBP and 12,600-13,900 yearBP. He believes that this north to south, easterly extinction path strongly supports the Blitzkrieg hypothesis. He admits that there are several issues to be resolved particularly the finding that the initial decline of mammals began earlier than estimated which, if it supports Martin's hypothesis, would indicate humans entered these regions earlier than other research indicated. However, the timing of the extinctions could be explained by the GTME. As explained earlier, the redistribution of polar ice mass caused a change in longitudinal surface gravity in a west-to-east manner. This would mean that surface gravity started increasing, from its lower level to a higher near-current level, in the Bering Strait first and then to the contiguous United States and then to South America. And, it also raises the question of whether the initial lower surface gravity in all of the three regions is what deterred human migration from entering these regions earlier.

3.1 THE AUSTRALIAN MEGAFAUNA EXTINCTIONS

If the GTME is valid as it applies to Quaternary extinctions then surface gravity on Australia had to be low for a period of time prior to about 50,000 yearBP when the megafauna there started to go extinct. If we look at a model of the globe we find that the area that is antipodal to Australia is the eastern edge of South America. Therefore, prior to about 50,000 yearBP, no megafauna should be found in that South American region. It is possible that dwarf megafauna existed there at that time and eventually increased in size in South America; the dwarfism would have been the result of higher surface gravity. The author of this book does not know how long the reduced surface gravity in the Australian region existed prior to 50,000 yearBP.

The research cited earlier concludes that megafaunal extinctions occur during the warming inter-glacial periods, the interstadials. Logically, we should look for such a period about the time of the start of the Australian extinctions. The following chart in FIG. 3-2 indicates that there was a significant cold stadial (GS-

18) about 62,000 yearBP through about 60,000 yearBP.

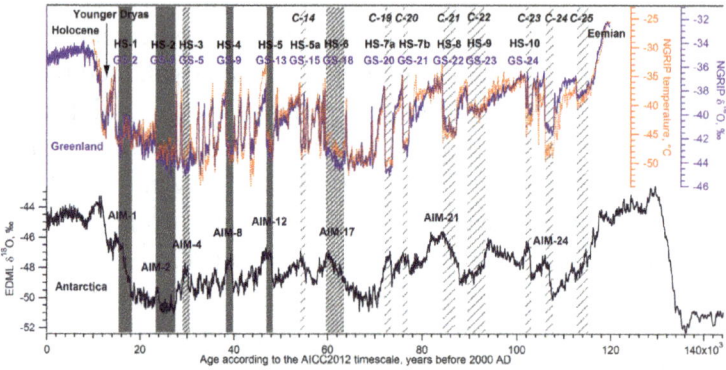

FIG. 3-2 Stadials (gray vertical bars) and interstadials courtesy Bozin et al.,2013;Veles et al., 2013

In FIG. 3-2 the orange dots represent temperature. The solid gray areas are major stadials of Laurentine origin. Dark gray hatch are major stadials of European origin. Light gray hatch are minor stadials registered in North Atlantic marine sediment cores.

The stadial GS-18 was followed by a warming interstadial of about 60,000 yearBP through 56,000 yearBP, a less significant brief stadial at 55,000 yearBP followed by an extended interstadial from about 54,000 yearBP through 48,000 yearBP. One or both of these warming periods may have been the cause of a significant polar ice melting resulting in the change in surface gravity on Australia.

24

Most of the megafauna of Australia consisted of marsupials. Diprotodon was the largest known herbivorous marsupial and resembled a rhinoceros. It is estimated that 19 genera, 50 species of mammals became extinct in Australia approximately 50,000-45,000 yearBP. Other mammals that went extinct were small ground sloths, giant kangaroos, tapirs, capybaras, small rhinoceroses and a large carnivorous cat.

The "overkill" hypothesis for Australia is weak. The absence of megafaunal bones, articulated or disarticulated skeletons, in close proximity to remnants of human activity, such as hunting tools, is an indication that the megafauna didn't become extinct due to human hunting although the blitzkrieg supporters claim the movement of the human hunters was so rapid that there was minimal chance of finding the remains of megafauna at campsites. A femur was found at Mammoth Cave with a notch in it but this could have resulted from, for example, a marsupial lion. Those who support the "overkill" hypothesis believe that the megafauna of Australia were not wary of the human hunters as they were in Africa and this naivety resulted in their extinction.

Some climate change proponents believe climate change alone was responsible for the extinctions but critics state that megafaunal species survived 2 million years of climatic oscillations, some with arid glacial periods prior to the extinctions. Oxygen and carbon isotopes of teeth indicate a sudden, drastic non-climate-related change in vegetation in the diet of surviving marsupial species. A sudden increase in regional surface gravitation can alter vegetation. For example, trees and plants with massive long limbs would obviously be affected. In fact, in describing the Carboniferous trees that were 160 or more feet in height with delicate fern-like leaves that sat on top of pencil-thin trunks Peter Ward and Joe Kirshvink wrote in their book[10] '*A New History Of Life*':

"*One of the strangest traits was a very shallow root system.*"

It's not so strange when one realizes that surface gravity,

according to the GTME, was lower at that time. The graph[1] of the latitudinal position of the center of mass (COM) of Pangea during the Carboniferous clearly shows it to be at a very high southern latitude (see References).

The time of arrival of humans in Australia is still being debated. A range of 44,200 to 71,500 yearBP has been proffered. Based upon this range, early humans could have coexisted with megafauna for an extended period of time eliminating the naivety factor or they arrived after the extinctions commenced or anything in between. There were no major local climatic changes in Australia nor evidence of human hunting during the Australian megafaunal extinctions. However, as with the megafaunal extinctions in North and South America, many scientists have attempted to attribute a combination of human hunting and climate change as the culprits in the extinctions. As stated throughout this book, the author believes surface gravity changes were responsible.

It is believed that early Australian Aborigines appear to have rapidly eliminated the megafauna of Tasmania about 41,000 yearBP following formation of a land bridge to Australia about 43,000 yearBP, as ice age sea levels declined, without using fire which implies that human hunting is the most likely cause of the extinctions. The problem with this hypothesis is that during this period the Wisconsin glacial polar ice was rapidly retreating which normally would result in higher eustatic sea levels. However, the GTME may be able to explain the lower sea levels near Tasmania at that time. As explained in Chapter 2, when the polar ice caps diminish in size in the longitudinal polar regions with the largest ice mass, not only does surface gravity increase in that longitudinal region (due to the offset core elements moving toward Earth-centricity) but sea level also lowers in that region, even lower than other places around the globe, especially antipodally.

As mentioned, most of the Australian megafauna were marsupials. If surface gravity there increased we would expect marsupials to be affected to a much greater degree than placental mammals of the same size. This is one of the reasons why the megafauna of Africa were less affected by increasing surface gravity than those of Australia. Also, African megafauna such as the elephant would be less vulnerable to the negative effects of increasing surface gravity compared to, for example, mastodons living in North America. With increasing surface gravity it takes a lot more food to fuel the physical activity of a mammal in a northern climate as well as the reduced availability of food in areas with seasonality. Increasing surface gravity would have a much greater effect on megafauna in northern regions.

3.2 THE NORTH AMERICAN MEGAFAUNAL EXTINCTIONS

Archeologist Jerry N. McDonald has done considerable research[5] regarding North American ecology during the terminal Pleistocene period. He notes that two-thirds of North America's large mammal genera along with reptilian and avian genera became extinct during the late Pleistocene (i.e., late Wisconsin) and early Holocene. The most familiar of the mammalian extinctions include the mammoth, mastodont, horse, sloth, camel and musk ox. McDonald analyzed many biological and environmental conditions that would have been prevalent during the period referenced and offered his opinion:

"Modern ecological and evolutionary theory predicts that most changes in North America's environment during the period of extinction should have resulted in an increase in megafaunal biomass and perhaps even diversity."

He questions whether the changing environmental conditions during this period were sufficient to cause the extinctions:

"Generally speaking larger-bodied mammals should be among the last categories of organisms to be brought to extinction by broad-spectrum environmental changes unless selection operated against some character or characteristics ubiquitous in, and peculiar to, the group."

His chart[5] shows that the biologically productive land area of North America increased by almost 70% between the late Wisconsin maximum (~18000 yearBP) and the middle Holocene (~7000 yearBP).

After supplying convincing data which questions the environmental causation for the extinctions he concludes:

"The appearance of human hunting as a new (or substantially expanded or improved) ecological process represented the only new, important, and ubiquitous factor introduced into North American selection regime during the very late Wisconsin."

The following statement by McDonald should be kept in mind when reading about the author of this book's new theory describing the reason for the megafaunal extinctions:

"Whatever the proximal cause or causes of the extinctions in North America, the phenomenon was continent-wide, not regional; it was selective principally against large mammals; and it resulted probably from some ultimate, ubiquitous causal factor. The magnitude and rate of faunal simplification further indicates that a significant change occurred in the continent's natural selection regime sufficient to render, quickly and decisively, much of the continent's megafauna 'unfit.' "

3.3 THE SOUTH AMERICAN MEGAFAUNAL EXTINCTIONS

Paul S. Martin[4] has written extensively about the megafaunal extinctions in South America. Most of what follows summarizes his research on this subject. Mr. Martin is well known for his Blitzkrieg hypothesis attributing the extinctions to rapid movement of early humans in North and South America. He notes that earlier extinctions in the Cenozoic were numerous. He writes:

"The Lujanian (~ 300kyearBP - 10k yearBP) extinctions stripped South America of
21 genera and 4 families of large edentates;
1 genera of large rodent;
3 genera of large carnivores;
2 last genera of the ungulate order Notoungulata;
4 genera of mastodonts and the order Proboscidea;
3 genera of horses and the family Equidae and
11 genera of artiodactyls including peccaries, camelide and deer."

Among the above megafauna are:
Arctotherium- short-faced bear- up to 3500 lbs.
Huppidion- similar to extant wild horses.
Macrauchenia- long-necked, long-limbed camel-like llama.
Pampatheros- similar to an armadillo but much larger- 450 lbs.
Smilodon- large saber-toothed cat- 220-620 lbs.
Stegomastodon- In appearance, similar to a mastodon but smaller- 5 tons.
Toxodon- large-hoofed mammal with rhinoceros-like appearance- up to 3300 lbs.

Comparing the scope of the extinctions in North and South America, Martin writes:
"The 46 Lujanian extinctions considerably exceed the large mammal extinctions of the late Pleistocene in North America. No other continent (i.e., other than S.A.) lost as many mammals in the late Pleistocene."

Reaffirming his belief of human causation of the extinctions, he writes:
"Human impact has been held accountable for the event, at least in part."

Note his following statement, which helps to support the author of this book's theory of the extinctions:
"Ground sloth extinction in South America may be slightly younger than in North America."

This is something one would expect if the early human hunters migrated from north to south, assuming they were the cause of the extinctions. However, if the increasing surface gravity was moving from west to east in North America at this time it would explain the delay of the ground sloth extinction in South America. He further states that:

"The date of human arrival in South America is unclear."

Some scientists believe early humans arrived in South America very early by boat, which if true would weaken Paul Martin's Blitzkrieg hypothesis.

Finally, Paul Martin sums up the basis for his Blitzkrieg hypothesis:

"As in North America, deposits rich in remains of late Pleistocene extinct fauna such as the tar seeps of Talara, Peru, the mineral springs of Araxa, Brazil, or the dung deposit of Mylodon Cave, Chile, have not yielded unimpeachable evidence of human artifacts in abundant association with extinct animals. For this reason I proposed a rapid overkill or blitzkrieg of the South American extinct fauna (Martin 1973)."

3.4 THE AFRICAN MEGAFAUNAL EXTINCTIONS

Often the point is made that the late Pleistocene megafaunal extinctions in Africa were minimal compared to other continents. The reason for this is frequently suggested to be the result of the lengthy period of time in which humans and megafauna coexisted eliminating the "naive factor." In other words the megafauna evolved a wariness of human predators. While there may be some validity to this hypothesis, the theory proposed by the author of this book, as explained in Chapter 2, can offer a different explanation.

Prior to the late Pleistocene there have been many examples of megafauna that went extinct since the beginning of Pleistocene Epoch (i.e., 2.60 myBP). Some of these extinctions are:

1. **Megantereon**- Comparable to a large jaguar but heavier (200-330 pounds) with elongated fangs. This carnivore lived in North America, Eurasia and Africa and went extinct in the middle Pleistocene.
2. **Homotherium**- Comparable to an African lion (330-500 pounds). This saber-tooth carnivore lived in North America, South America, Eurasia and Africa. It first became extinct in Africa 1.5 myBP and in Eurasia 30,000 yearBP.
3. **Deinotherium**- A prehistoric elephant much larger than the existing African elephant. It had a shorter trunk and downward curving tusks attached to the lower jaw. It weighed up to 14 tons and was 13 feet tall at the shoulders. The youngest known Deinotherium is from the early Pleistocene of about 1 million yearBP.
4. **Dinofelis**- Comparable to a jaguar in size, this saber- toothed carnivore was found in Europe, Asia, Africa and North America. It became extinct in the early Pleistocene of about 1.2 million yearBP.

Richard G. Klein, paleoanthropologist, provides a background of the megafaunal extinctions in Africa[7]. Excluding the extinction of Kirchberg's rhinoceros, a special case, it appears that no mammals became extinct in Africa during the Upper Pleistocene (130,000 yearBP to 12,000 yearBP) until the terminal Pleistocene:

"As a complete explanation for the extinctions, however, terminal Pleistocene/early Holocene environmental change is insufficient, since the extinct species clearly survived earlier periods of similar change, occurring from at least the Middle Pleistocene on."

Klein suggests long-term changes in the way Stone Age people interacted with animals could be key to the extinctions. His conclusion is that Late Stone Age (LSA) people were much better hunters of large mammals than Middle Stone Age (MSA) people and this could explain the extirpation of megafauna at the terminal Pleistocene. Klein's research provided a table listing a considerable number of large mammal species that became extinct in Africa during the Lower and Middle Pleistocene (i.e.,

1.8 Ma to 130,000 yearBP). He attributes these extinctions to:
(1) the failure of species to adapt to changes in climate or physical environments, or
(2) their failure to compete successfully in biotic communities that were constantly changing as a result of evolution in local species and the immigration of foreign ones.

Klein notes that:
"The pace and extent of extinctions that occurred during the Pleistocene and Holocene in Africa seem broadly comparable to the extent of extinctions that occurred during the same interval in Eurasia."

He further notes that:
"On neither continent is there evidence for a major wave of extinctions, wiping out a very large portion of the existing fauna in a relatively brief time."

The above two statements provide strong support for the theory of this book's author. Africa and Eurasia are in the same longitudinal region, therefore they would experience about the same change in gravitation. If surface gravity gradually increased over time in the Eurasia/African longitudinal region, fauna could adapt through evolutionary change during the transitional period or migrate east or west to an area where surface gravity didn't increase as much. At the terminal Pleistocene the surface gravitational changes occurred too quickly and intensely for any adaption to occur.

Regarding the late Pleistocene and early Holocene megafaunal extinctions Klein concludes that they can be attributed to the presence of anatomically modern people with superior hunting skills than those people of the Lower and Middle Pleistocene periods, coupled with dramatic climate changes.

3.6 DWARFING

On the subject of Pleistocene dwarfing of megafauna senior research scientist Larry G. Marshall[6] writes:

"The knowledge that extinction and dwarfing are concurrent processes may provide new insight into extinction phenomena. If we can identify mechanisms causing a decrease in body size, we can, so to speak, approach extinction 'through the back door' assuming that both extinction and dwarfing are indeed linked to a common causal factor(s). An explanation of the dwarfing may then be extrapolated to explain the extinctions."

The author of this book agrees with Marshall's assessment of the relationship of dwarfing and extinction and believes that the linkage between the two are a direct result of changes in surface gravitation.

Paleobiologist R. Dale Guthrie[8] [2003] pointed out the decline in body size and eventual extinction of horses in Alaska. However, he attributed this to vegetation changes by a warming climate, decreasing access to optimal foods.

Most of the literature on dwarfing deals with one aspect of dwarfing which is primarily associated with the insular island reduction of the size of fauna, primarily mammals. What appears to happen when mammals become stranded on islands, particularly when carnivores are absent, is the gradual reduction in size of the largest mammals. This phenomena has been studied on Mediterranean and Indonesian islands.

The dichotomy of opinion on the cause of the dwarfing falls along the same lines regarding megafaunal extinctions. Some scientists believe climate and environmental factors favor fauna that reduce their energy requirements conserving resources, thereby reducing the quantity of plant life needed for consumption. Other scientists believe human hunters are more likely to kill the largest members of their favorite prey resulting in the gradual reduction in physical size of future offspring of the surviving prey.

A new study[3] suggest that neither one of the two primary hypotheses is valid. '*The Island Rule: Made To Be Broken*' casts doubt on the hypothesis described above. The authors conclude:

"*Size evolution on islands is often thought to be tightly related to characteristics of the islands and their mammalian faunas, such as island area, isolation and the presence or absence of carnivores (Heaney 1978; Michaux et al., 2002). We found little evidence that these factors have a consistent influence on body size evolution.*"

This study is based upon extant fauna that would obviously not be subject to variations in surface gravity strength. We know that mammalian dwarfism existed at various times within the Pleistocene Epoch on both island and non-island environments. Referring to dwarfism of mastodons the authors indirectly support the possibility that there were other factors in the past that might explain the Pleistocene dwarfism:

"*We did not analyze these most extreme cases and it maybe that comparing more distantly related taxa would reveal stronger patterns.*"

CHAPTER 4
THE NEANDERTHALS

Neanderthals lived in Europe from about 400,000 yearBP to about 40,000 yearBP. Hypotheses concerning the extinction of Neanderthals (Homo neanderthalensis) are as diverse as those attempting to explain the Pleistocene megafaunal extinctions. These include climate change, competition from the newly arrived Early Modern Humans (EMH) and the volcanic eruption near Naples, Italy at the Phlegraean Fields (Campi Flegrei) of approximately 39,000 yearBP. The missing hypothesis is the surface gravitation change described in the GTME, which the author believes is the valid explanation for the extinctions of both the Neanderthals and the megafauna.

Just as the Ice Age megafauna attained unusual physical size, so did the Neanderthals. In a lowered gravitational environment, fauna whether human or not, can develop body structures far different from that of extant fauna. Some fauna will reach larger proportions constrained by the environment and the ability to "make a living" including the ability to discourage predators. Neanderthals, evolved in a much colder environment than their southern contemporaries in Africa, although both were in the same longitudinal area experiencing lower surface gravity, according to the GTME. The Neanderthals developed bodies with a larger mass to surface area ratio in order to retain heat and lower surface gravity enhanced this physical development. In Africa, the opposite was happening. Most of Africa, being in the same longitudinal range as Europe, experienced much lower surface gravity and higher temperatures allowing human bodies to have much higher surface area to mass characteristics, e.g., longer arms and legs and taller. Neanderthals' stockier body plan forced them to become ambush hunters; they were not fleet-footed hunters. Their right arm bones were much thicker than their left arms suggesting they relied on the right arm to deliver the most thrust when using a spear while hunting in ambush mode. Their protruding facial structure allowed the air they breathed to be pre-heated and pre-humidified before entry into the lungs. Their skull and brain were larger than ours.

If the Neanderthal extinction was caused by increasing surface gravity we would expect the extinction to be gradual and start during an interstadial warming period (see Fig. 3-2) when polar ice in the longitudinal region shared by Europe (and Africa) started to melt and the melt-off water was distributed to a lower latitude. Because the Neanderthals occupied the entire European continent and part of Asia, we would expect the extinctions to have occurred in different parts of this region at different times as surface gravity varied depending upon the polar ice configuration. And, if anatomically modern humans were absent during the Neanderthal extinction period in locations known to have been occupied by the Neanderthals this would bolster the gravitational theory. This absence of anatomically modern humans was the situation in parts of Europe.

An article in the publication Nature (08/14) concluded that the European Neanderthals disappeared 41,000 yearBP to 39,000 yearBP. However, recent research[15] at the El Salt site in Spain indicates that the Neanderthals disappeared there 45,000 yearBP. Crissto Hernandez said, regarding the last Neanderthals:
"A progressive weakening of the population, or rather, not towards an abrupt end, but a gradual one, which must have been drawn out over several millennia, during which the human groups dwindled in number."

Anatomically modern humans had no role in the disappearance because they did not enter this region of Spain which was shielded by the Pyrenees mountain range until much later. The GTME would interpret this pattern of extinctions in Europe as the eastward movement of increasing surface gravity across Europe and beyond as a result of the melting/redistribution of the polar ice caps such that ice caps in the same longitudinal area as Spain were retreating and/or antipodal ice caps were increasing in size.

In any Neanderthal region within Eurasia megafauna also existed. It would be logical to assume that in a region with lower surface gravity the largest megafauna in that region would become extinct prior to the Neanderthals in that region as surface gravity increased. They would both be negatively affected by increasing surface gravity but the Neanderthals, being significantly smaller, wouldn't be affected as quickly by the surface gravity increasing. The Neanderthals might have been able to target smaller prey and change their diet habits but they would eventually succumb to increasing surface gravity. The number of archeological sites is probably limited for verification purposes but it would be expected, if surface gravity was gradually increasing, that the last remaining Neanderthals in any region would be physically smaller than earlier members.

The discovery of ancient human teeth in a Chinese cave indicates that modern humans left Africa more than 80,000 yearBP. Yet they didn't arrive in Europe until about 45,000yearBP. Researchers suggest something prevented H. sapiens from entering Europe for tens of thousands of years.

Anthropologist Maria Martinon-Torres[17] of the University College London believes the Neanderthals prevented the H. sapiens entry into Europe. She believes that when they migrated north from Africa and reached the Levant they encountered the Neanderthals who blocked their northern destination forcing them to migrate east eventually reaching China.

The reader knows what the author of this book believes was the barrier that H. sapiens encountered when they left Africa, which was a lower surface gravity in Europe.

FIG. 4-1 Irish elk went extinct 10,000 yearBP, American
Museum of Natural History

CHAPTER 5
VOLCANISM, MEGAFAUNA AND NEANDERTHAL EXTINCTION

One hypothesis attributes the eruption of the Campanian Ignimbrite (CI) eruption from the Phlegraean Fields near Naples, Italy as the cause of extinction of the Neanderthals and possibly some of the megafauna. The eruption took place about 39,400 yearBP which is soon after the period (45,000 yearBP to 40,000 yearBP) that is most commonly cited for the Neanderthal extinctions. Based on this timing many critics reject this hypothesis because the eruptions appear to post-date the extinctions.

One of the underlying concepts of the GTME is that flood basalt volcanism is the direct result of one or more of the core elements moving toward Earth-centricity, primarily the inner core within the outer core. This concept is well supported by the fact that most, if not all, mass extinctions are accompanied by this type of volcanic eruption. All of the Big Five Mass Extinctions and other significant extinctions are associated with flood basalt volcanic eruptions[2]. Many scientists believe that those volcanic eruptions are the primary cause of the extinctions by altering the environment. The author of this book believes that although they may have contributed to and extended the duration of the extinctions, surface gravity changes preceded each of the volcanic eruptions and was the primary extinction mechanism. Flood basalt volcanic eruptions are believed to initiate at the Earth's core-mantle boundary unlike other volcanic eruptions that result from tectonic plates being subducted beneath other plates. Because it would take some time for a lava flow to initiate at the core-mantle boundary and eventually reach the surface, it would be expected that there would be a delay between the core movement (causing an increase in regional surface gravity) and the eventual eruptions at the Earth's surface, according to the GTME. This accounts for the Campanian Ignimbrite volcanic eruptions occurring a thousand or more years after the Neanderthals' demise due to a pulse of increasing surface gravity. In the regions that were shared by both megafauna and Neanderthals, both would eventually be driven to extinction by the same non-human cause, gravitational change.

The GTME raises an important question regarding the timing and entry of Early Modern Humans (EMHs) into regions that were occupied by Neanderthals. Could the EMHs function in a lower surface gravity region? At first glance one might assume that humans living in normal surface gravity as it exists today would have no problem adjusting to lower gravity. However, this is probably not the case. Even pragmatic tasks such as how a spear is thrown or how an arrow's path is altered by lower surface gravity would be problematic. A boat would float higher in water in an area with lower surface gravity. The EMH interlopers more likely entered these regions as surface gravity was increasing to near current strength so that they could coexist with the Neanderthals during this transitional period. This may be the reason they didn't enter those areas much sooner. And, the arrival of the EMHs during this transitional period may be the reason why the hypothesis that Neanderthals went extinct because they were outcompeted or wiped out by the EMHs has developed.

CHAPTER 6
THE HOBBIT (Homo floresiensis) AND THE GTME

These diminutive early inhabitants of the island of Flores in Indonesia were approximately three and one half foot in height. Scientists have concluded that their size was a result of insular island dwarfism. While that is a possibility, their diminished size could very well be a result of increased surface gravity. Skeletal remains and tools have been found for the period 94,000 yearBP to 13,000 yearBP.

When the first skeletal remains were found it was believed that H. floresiensis was a deformed early human, possibly suffering from microcephaly or other ailments. After significant analysis this view has been rejected.

Recent research[15] (11/2015, published in PLOS ONE) suggests:

"The Hobbit's teeth and skull were most similar to those of Java Man. What is extraordinary is that **this means the Hobbit, over a relatively short period of time, shrank in stature from about 1.5 meters tall to just over a meter.**"

There are physical characteristics of the Hobbit that might be explained by an era of higher surface gravity on that part of the globe rather than insular island dwarfism. From Wikipedia:

"The feet of H. floresiensis were unusually flat and unusually long in relation with the rest of the body. As a result, when walking, it would have to bend its knees further back than modern people do. This forced the gait to be high stepped and walking speed to be low."

Also, the claim that Homo floresiensis was a previously unidentified hominid was supported by, among other characteristics:

"...the thickness of the leg bones."

Another research paper[11] described the same physical characteristics of other fauna on Flores:
"Many morphological characteristics, such as short stature, robust lower limbs, large feet and a small brain are shared by a number of other insular mammal taxa from Flores and other islands."

The gait of the Hobbit was less efficient than that of modern humans. The foot had a flat arch that lacked the spring-like structure that could store and release energy when running. They had wide leg bones compared to the length of the leg. The females had wider pelvises than H. sapien females. Why would the Hobbit have these unusual physical characteristics but the living pygmy didn't develop the same characteristics?

Based on the physical characteristics listed above one can conclude that this small hominid would not be a swift moving hunter which one might expect but rather a plodding ambush predator, paradoxically the same modus operandi attributed to Neanderthals. The megafauna that H. floresiensis hunted were also dwarf-sized including a dwarf-sized mastodon, Stegodon florensis insularis. Was this due to insular island dwarfism or higher surface gravity? The author of this book believes it was the latter.

If the physical size and leg and feet characteristics of H. floresiensis are attributable to higher surface gravity then we should be able to match its characteristics with those fauna that lived in approximately the same longitude at the same time. The previous quote from the referenced research paper[11] appears to support that, e.g., the Sulawesi Island of Indonesia. Pleistocene island dwarfism also occurred in the Mediterranean region in places such as Cyprus, the Cyclades and Rhodes. The longitude of these locations are within the same range of the region occupied by Neanderthals. Therefore, if higher surface gravity was the cause of the dwarfism it would be expected that the dwarfism would commence while the Neanderthals were becoming extinct during the upper Pleistocene. Similar patterns of dwarfism could have happened during earlier periods.

The remains of 12 individuals found were dated from the period 38,000 yearBP to 13,000 yearBP at Liang Bua Cave but fewer older remains have been found. The exact dates are not clear except for the female, identified as LB1, of about 18,000 yearBP.

In Australia, megafaunal extinctions were taking place from about 50,000 yearBP or a little before. The GTME attributes this to a period of increasing surface gravity at that time in that region. Since Australia is in the same longitudinal region as Indonesia an increasing surface gravity in that shared region would be consistent with the presence of the Hobbit at that time.

The question arises about how H. floresiensis could function during the pre-50,000 yearBP period when Australia, according to the GTME, experienced lower surface gravity. Not finding any Hobbit remains in that period could signify that they weren't there then. Could they have migrated to other areas long before when surface gravity started lowering and then returned when surface gravity increased? Or, could they have adjusted to the lowering surface gravity without any significant physical change? It seems plausible that, unlike the Neanderthals who would be severely affected by an increase in surface gravity and energy requirements, the Hobbit might have been able to adjust to a lower surface gravity because its energy requirements would have been lowered.

The date of disappearance of H. floresiensis has been given as 11,000 yearBP. Another eruption of flood basalt volcanism at the Phlegraean Fields in Italy occurred about 12,000 yearBP. The GTME links this volcanism with core element movement toward Earth-centricity. In Chapter 2 of this book a description of the GTME describes how flood basalt volcanism occurs whenever the Earth's core(s) move rapidly toward Earth-centricity. The core movement would also have caused a rapid lowering of surface gravity on Flores and increasing antipodally in North and South America, the speed of which may have been too fast for H. floresiensis and the megafauna in that region to adjust to. This would happen if the polar ice caps, rather than redistributing longitudinally, rapidly melted causing surface gravity globally to approach the current level, which appears to have happened because extinctions occurred at about the same time around the globe.

CHAPTER 7
GEOMAGNETIC REVERSALS AND THE GTME

The Laschamp geomagnetic field reversal found in the sediments of the Black Sea indicates that a reversal of the Earth's magnetic field took place about 41,000 yearBP. Coincidently, the Campanian Ignimbrite (CI), the largest flood basalt volcanic eruption in the Northern Hemisphere occurred at the Phlegraean Fields near Naples Italy 39,400 yearBP. As mentioned previously, some scientists attribute the extinction of the Neanderthals to the environmental effects of this volcanic eruption.

It was also noted that the author of this book believes that the GTME can explain the coincidence of the extinction of the Neanderthals and the volcanic eruptions and the geomagnetic reversal. FIG. 7-1 illustrates (when Pangea existed) how the off-set core elements, when moving back toward Earth-centricity reduced the Earth's surface gravity gradient around the globe. In other words, surface gravity increased where it was lower and lowered antipodally where it was higher.

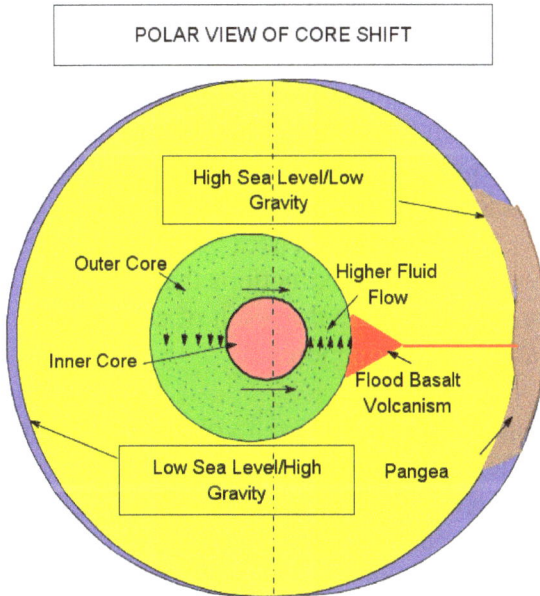

FIG. 7-1 Core shift, surface gravity chg., sea level chg. & volcanism

FIG. 7-1 illustrates why the Earth's magnetic field reversal occurred. When the inner core is not centrally disposed within the outer core, the molten iron liquid's flow is disturbed so that the liquid's velocity is significantly different within the two halves of the hemispheres of the outer core. The hemispheres are not the north and south hemispheres but the east and west hemispheres. The flow in the narrower space must be faster than on the opposing hemisphere of the outer core. This alters the magnetic field causing it to reverse polarity. The faster flow of the molten liquid in the narrower region delivers additional thermal energy to the adjacent core-mantle boundary initiating a pulse of the flood basalt lava flow which eventually reaches the surface. The lava would reach the Earth's surface well after the extinctions were initiated which seems to agree with the Neanderthals extinction (i.e., 45,000 yearBP to 40,000 yearBP) and the volcanic eruption of about 39,400 yearBP.

A link between geomagnetic field reversals and extinctions has been hypothesized several times in the past. One hypothesis states that atmospheric oxygen level and geomagnetic field intensity affect the evolution of life. The research paper[18] suggests that oxygen can be lost when the protective magnetic field of the Earth weakens during geomagnetic reversals resulting in periods of extinction due to the drop in ambient oxygen levels. Based on the GTME, the geomagnetic field reversal is not the cause of extinctions that are nearly coeval with the extinction events but are the result of core element(s) movement, as are the extinctions.

The coincidence of the flood basalt volcanic eruption and the geomagnetic field reversal is not a random coincidence. During the following three periods of the Pleistocene there was the coincidence of extinctions, flood basalt volcanic eruptions and geomagnetic field reversal:

1. The terminal Pleistocene (12,000 yearBP) had the flood basalt eruption at the Phlegraean Fields, the massive megafaunal extinctions and the Gothenburg geomagnetic field reversal.

2. The upper Pleistocene (45,000 yearBP to 39,400 yearBP) had the flood basalt eruptions at the Phlegraean Fields again, the megafaunal and Neanderthal extinctions and the Laschamp geomagnetic field reversal.

3. The lower Pleistocene (780,000 yearBP) had the flood basalt eruptions at Mount Taveuni, Fiji, the significant deep-sea benthic foraminifera extinctions at Challenger Plateau of the East Tasman Sea (east of Australia) and the Brunhes-Matuyama geomagnetic field reversal. The physical proximity of extinctions and volcanic eruptions at this time, according to the GTME, indicates that this longitudinal region had lower surface gravity up to that time and the antipodal area, the westernmost part of Africa and Europe, had higher than present surface gravity.

The only theory that can explain the coincidence of all three phenomena occurring at the same time is the Gravity Theory of Mass Extinction.

FIG. 7-2 Lestodon-ground sloth of South America went extinct about 30,000 YearBP, American Museum of Natural History

CHAPTER 8
SUMMING IT UP

The **Gravity Theory of Mass Extinction (GTME)** is based on the fundamental belief that the Earth's core elements (i.e., the inner core, outer core and the densest part of the lower mantle) can move away from their Earth-centric position and have done so significantly in the past. This offsetting phenomena can only happen when mass, especially a large mass, on the Earth's surface moves latitudinally and when this occurs, by definition, surface gravity around the globe must be non-uniform, unlike the way it is today. At minimum, the inner core would move without the other 2 core elements moving off-center although all three moved off-center during the Carboniferous (and before) and the Permian through the Cretaceous (and later).

The above core element phenomena, based on the GTME, is responsible for:
1. All major mass extinctions
2. All major flood basalt volcanic eruptions
3. All geomagnetic field reversals
4. The Earth's secular variation
5. Most non-eustatic sea level changes

In the distant past when supercontinents existed, as in the case of Pangea or earlier with Laurasia and Gondwana, surface gravity on Earth was non-uniform due to the latitudinal imbalance of continental mass. This accounts for the enormous size of some taxa including the dragonfly (Meyaneuropsis permiana) of the late Carboniferous with a two-foot wingspan and the sauropods of the Mesozoic. When the center of mass of the cumulative continental mass moved toward lower latitudes the core elements moved toward Earth-centricity causing increasing surface gravity in regions that had lower surface gravity, lowering sea levels near those regions, initiating flood basalt volcanic eruptions and causing extinctions, due primarily to increasing surface gravity. However, in the distant past there was an additional factor in the extinctions, the disassociation of methane from the hydrates at the bottom of the sea. This release of methane was caused by the combination of three factors:

1. The lower surface gravity (i.e., lower than today's level). Lower surface gravity reduces the water pressure per unit depth.
2. Lower sea level (i.e., lower than today's level).
3. Relatively warm temperature of the seawater.

Just about every mass extinction was characterized by a carbon isotope excursion. Many scientists attribute this excursion to the volcanic eruptions. The volcanism enhanced this phenomenon but was not sufficient to account for the large negative swings in delta 13c, leaving the disassociation of methane from the methane hydrates at the bottom of the sea responsible. As stated above, the GTME attributes the methane disassociation from the methane hydrates at the bottom of the sea from warm sea water and low water pressure.

The same core element offsetting phenomena occurred more recently during the ice ages. Instead of tectonic plates moving to a higher latitude, ocean water was transferred to the polar region and converted to ice causing one or more of the core elements to move off-center. As the polar ice configuration varied so did the corresponding longitudinal surface areas vary in their surface gravity magnitude. The lowest surface gravity existed in the longitudinal region with the most polar ice because the core elements moved away from that longitude within the equatorial plane and the antipodal longitudinal areas had higher surface gravity, relative to today's surface gravity. During the interstadials when the polar ice mass diminished in a longitudinal area, the core elements moved back toward Earth-centricity toward the same area causing surface gravity there to increase. This is the primary reason why the extinctions during the ice ages occurred on the continents at different times and always during the interstadials when polar ice melted. These extinctions occurred independent of the local climate changes, something which has confounded the "Overchill" supporters and has buoyed the "Overkill" enthusiasts. As stated earlier, the former group is, however indirectly, supported by the GTME because the interstadials set in motion the core elements' movement although the "Overchill" group hasn't given the true reason for the extinctions, i.e., surface gravity change.

The GTME also explains why the Neanderthals were physically different from the early modern humans that migrated from Africa about 60,000 yearBP and why the Neanderthals went extinct. They evolved in a low gravity environment in Eurasia. It is not a coincidence that they disappeared near the time of the massive flood basalt volcanism of the Phlegraean Fields which was accompanied by the Gothenburg geomagnetic field reversal. The relationship between massive flood basalt volcanism and mass extinction was explained in Chapter 2 and the relationship between mass extinction, flood basalt volcanic eruptions and geomagnetic field reversals was described in Chapter 7. They are all related to the offsetting movement of the Earth's core elements. The author of this book also believes that the offsetting movement of the core elements is responsible for the Earth's magnetic field secular variation, the movement of the magnetic north and south poles. This implies that the changes in the configuration of polar ice is primarily responsible for changes in secular variation during the last few million years.

The Hobbit (H. floresiensis) existed primarily in the region with higher surface gravity within the Indonesian area, while the Neanderthals evolved in a low surface gravity environment in Eurasia. This would account for the diminutive size and other physical characteristics of H. floresiensis.

The best explanation that this author can give for the coincidence of flood basalt volcanism and geomagnetic field reversal at the time of extinctions as well as the continuous magnetic secular variation is that when the Earth's inner core becomes noncentric within the equatorial plane of the outer core the flow of liquid iron around the inner core is no longer uniform because the space between the inner core and the perimeter of the outer core will be a nonuniform width. This narrower region will be on the opposite side of the outer core from which the inner core was radially displaced. The liquid iron within the outer core will then be forced to flow at a higher rate in this region transferring much more heat to the nearby mantle initiating flood basalt volcanism. This nonuniform flow of molten iron also causes a disturbance to the Earth's magnetic field which can rapidly cause the reversal of the magnetic field.

The current conflict between the "Overchill" and "Overkill" hypothesists is reminiscent of the longstanding conflict between the asteroid impact and volcanism supporters concerning their hypotheses on the causation of the Cretaceous-Triassic Extinction. What is interesting is the recent attempt to meld the two hypotheses (for both the Ice Age extinctions and the end-Cretaceous extinctions) together resulting in a complementary explanation for the respective extinctions. The author of this book believes both extinctions can be explained by the GTME.

Present day climatologists are constantly warning us about anthropological causes of climate warming, which they should do. There is a long list of negative effects which we face if we don't solve this problem soon. However, there is a potential problem that could arise based on the GTME. If another ice age occurs and it is as intense as those of the past there will be massive extinctions. Clearly the higher latitudes will be uninhabitable as they were during prior stadials. But the non-obvious catastrophe will be the changes to surface gravitation around the globe. People who would have been living in regions that would eventually have either lower or higher surface gravitation than that of today would have to gradually migrate to other regions as well as migrate the livestock they own. And, the location of these regions would change as the polar ice redistributed longitudinally. How would they, or could they, evolve if they remained in the regions with altered surface gravitation? This massive reduction of livable land would create major conflicts. Since flora would also be affected by a changing surface gravitation, this would have further negative effects. One can only conclude that we must limit global warming but not totally eliminate it in order to prevent another ice age.

Also, as has been explained in this book, large flood basalt volcanoes are the result of the Earth's core elements moving back toward Earth-centricity during an interstadial. This significant volcanism would cause an additional catastrophe to those listed above when the next ice age ends. This is another reason to avoid a future ice age. When it comes to fighting global warming the old adage about "Everything in moderation" is applicable here.

REFERENCES

1. The following chart illustrates the relationship between the latitude of Pangea's center of mass, sea level and mass extinction for the last 300Ma. The continental symmetry chart is derived from the research paper '*Plate tectonics may control geomagnetic reversal frequency*' by F. Petrelis, J. Besse and J.-P. Valet (10/11/2011); Geophysical Research Letters, Vol 38, L19303

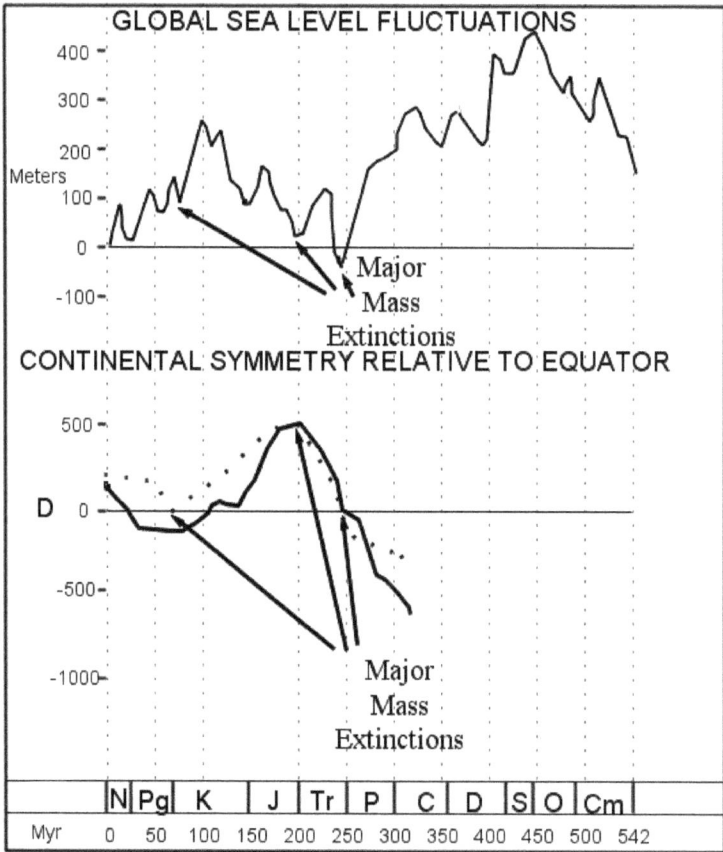

The chart above provides evidence that when Pangea's center of mass moved rapidly toward the equator mass extinctions occurred. The dotted line represents Christopher Scotese's view of Pangea's latitudinal movement and appears to be the more accurate one for the K-Pg extinction. See www.scotese.com for maps of ancient Earth.

2. The following chart illustrates the coincidence of mass extinctions and flood basalt volcanism from ~500Ma to ~15Ma. The GTME posits that all of these episodes of flood basalt volcanism are created by the movement of the Earth's core elements toward Earth-centricity.

The cause of the flood basalt volcanism events of the ice ages are related to the extinctions of the distant ones shown in this chart in that the same mechanism initiated them, i.e., core movement.

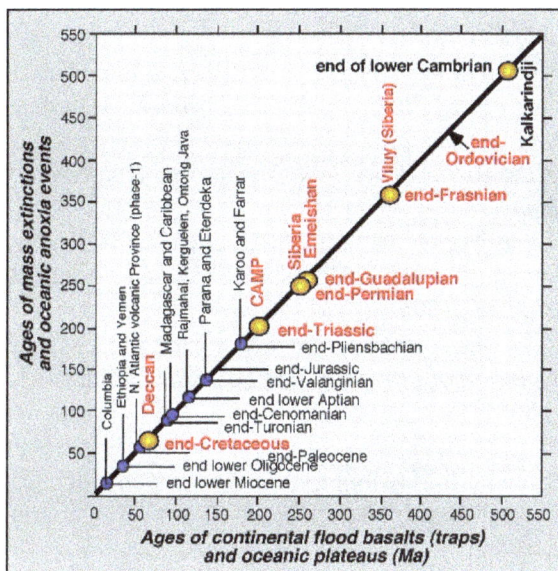

3. *The island rule: made to be broken?* Shai Meiri, Natalie Cooper, Andy Purvis; Proceedings of the Royal Society 2007.
Quaternary Extinctions: A Prehistoric Revolution
Paul S. Martin, Richard G. Klein, Editors ISBN 0-8165-1100-4, 1984
4. *Quaternary Extinctions*, Chapter 17 *Prehistoric Overkill: The Global Model*
5. *Quaternary Extinctions*, Chapter 18 *The Reordered North American Selection Regime and Late Quaternary Megafaunal Extinctions*
6. *Quaternary Extinctions*, Chapter 36 *Who Killed Cock Robin? An Investigation of the Extinction Controversy*
7. *Quaternary Extinctions*, Chapter 25 *Mammalian Extinctions and Stone Age People*
8. *Quaternary Extinctions*, Chapter 13 *Mosaics, Allelochemics, and Nutrients: An ecological Theory of Late Pleistocene Megafaunal Extinctions*
9. *Abrupt warming events drove Late Pleistocene Holarctic megafaunal turnover*
Alan Cooper, Chris Turney, Konrad A. Hughen, Barry W. Berok, H. Gregory McDonald, Corey J.A. Bradshaw; Science 07 Aug 2015: Vol. 349 Issue 6248 pp. 602-606
10. *A New History of Life: The Radical New Discoveries about the Origins and Evolution of Life on Earth*; Peter Ward and Joe Kirshvink ISBN-13 978-1608199075 3/10/15
11. *The fellowship of the hobbit: the fauna surrounding Homo floresiensis;* Hanneke J. M. Meijer, Lars W. Vanden Hoek Ostende, Gert D. Van den Bergh & John de Vos; Journal of Biogeography (2010) 37, 995-1006
12. *Chaos killed the dinosaurs*: Nature News; published online 28 June 2001, Nature doi: 10.1038/news010628-15
13. *Quaternary Extinctions*, Chapter 9 *Ten Million Years of Mammal Extinctions*
14. *Test of Martin's overkill hypothesis using radiocarbon dates on extinct megafauna*; Surovell, Pelton, Anderson-Sprecher, Meyers; Dept. Of Anthropology, Univ. Of Wyoming 8/5/15
15. *New evidence of early Neanderthal disappearance in the Iberian Peninsula*, Galvan, Hernandez, Mallol, Mercier, Sistiaga, Soler; Journal of Human Evolution, 2014; 75: 16 DOI; 10.1016/j.jhevol.2014.06.002
16. *Unique Dental Morphology of Homo floresiensis and Its Evolutionary Implications*, Kaifu, Kono, Sutikna, Saptomo, Due Awe; PLOS/one,11/18/15; DOI:10.1371/journal.pone.0141614

17. *The earliest unequivocally modern humans in southern China,*
Wu Liu, Maria Martinon-Torres et al., Nature
doi:10.1038/nature15696
18. *Oxygen escape from the Earth during geomagnetic reversals,* Yong
Wei, et al.; Earth and Planetary Science Letters Volume 394, 15 May
2014, pages 94-96
19. *The Gravity Theory Of Mass Extinction*, 3rd Edition, John
Stojanowski, ISBN 97809819221-4-0

INDEX

www.ingramcontent.com/pod-product-compliance
Lightning Source LLC
Chambersburg PA
CBHW071123210326
41519CB00020B/6396